KITCHEA Ⅱ

KITCHEA II

출판사 : 도시계획
발행인 : 고경철
주　소 : 경기도 안양시 만안구 안양동 13
전　화 : 031-473-2213
가　격 : ₩68,000원

013

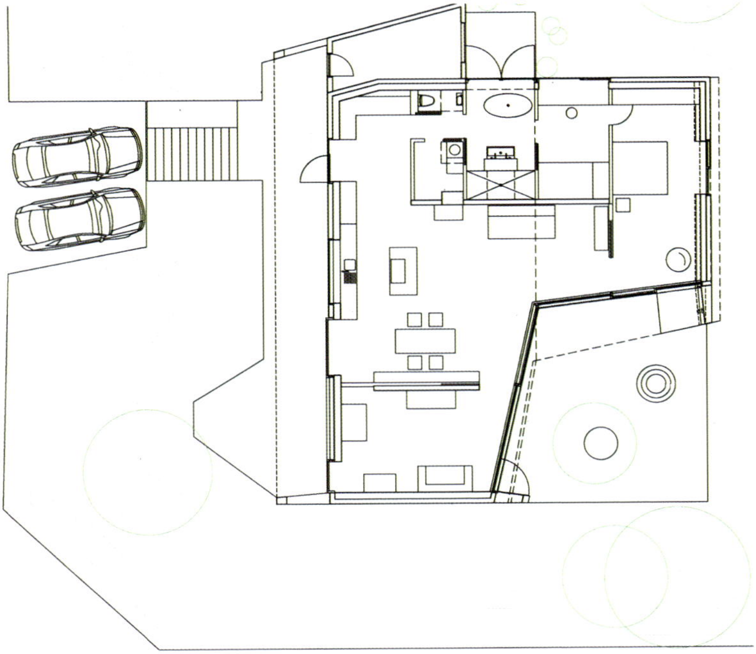

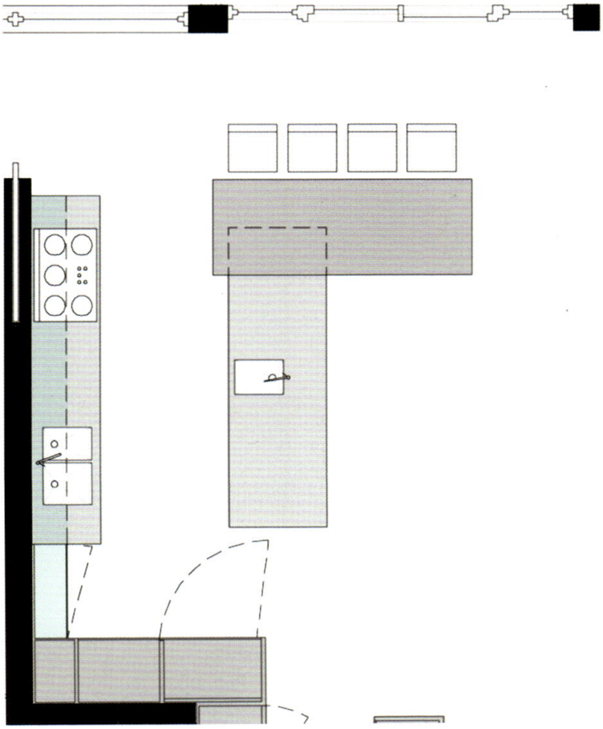

Omelette
tablets 8mg
10 tablets
HirstDamien

Wallpaper*

ROTHKO

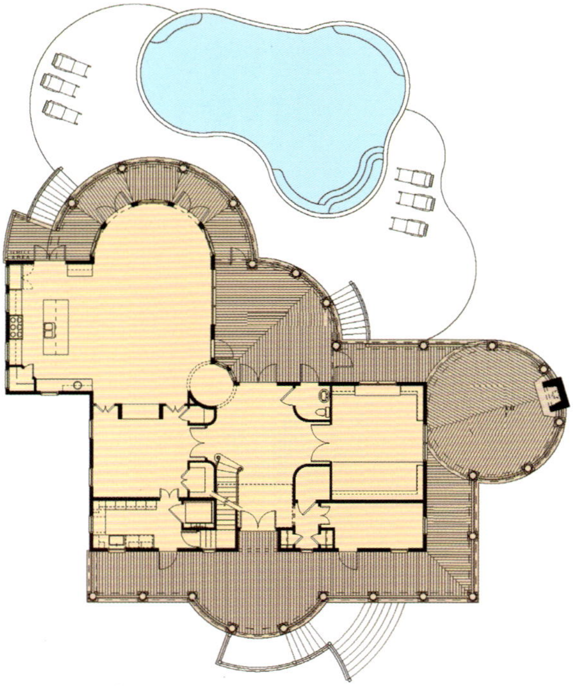

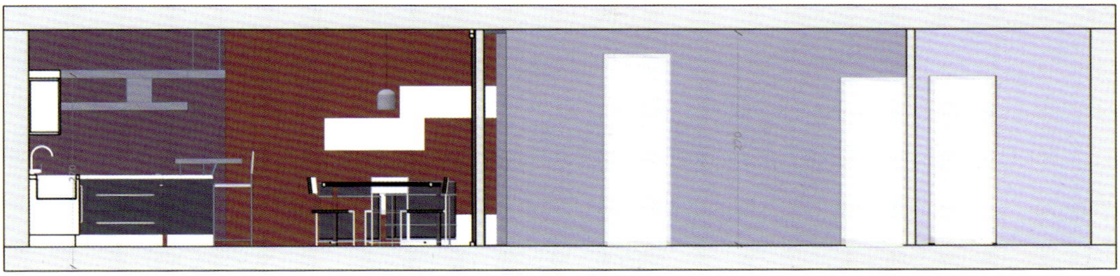
270

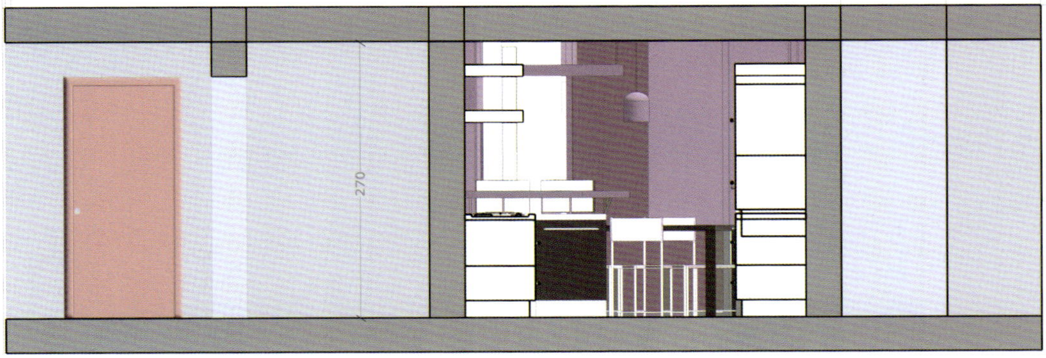
270

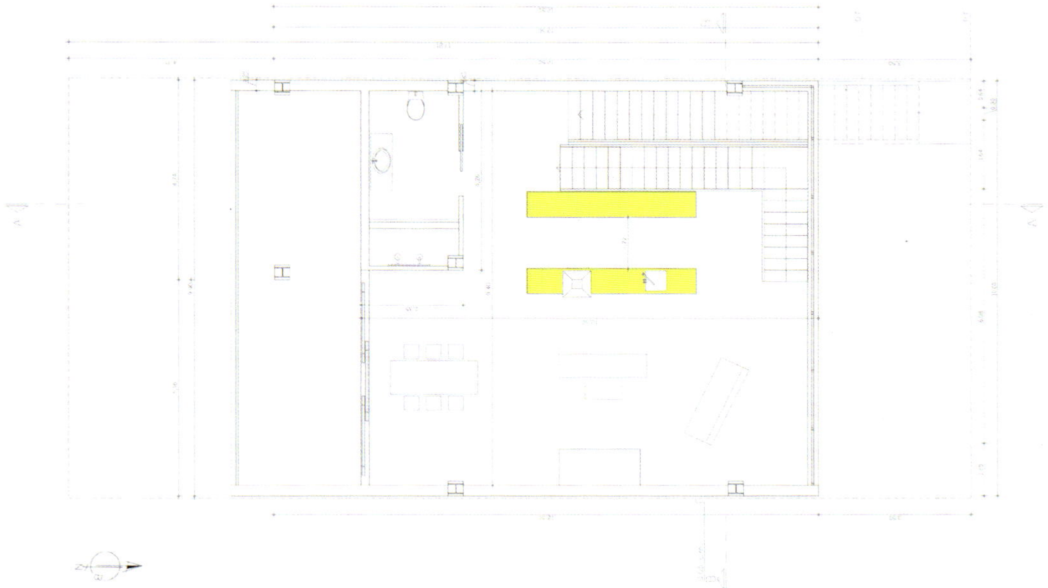

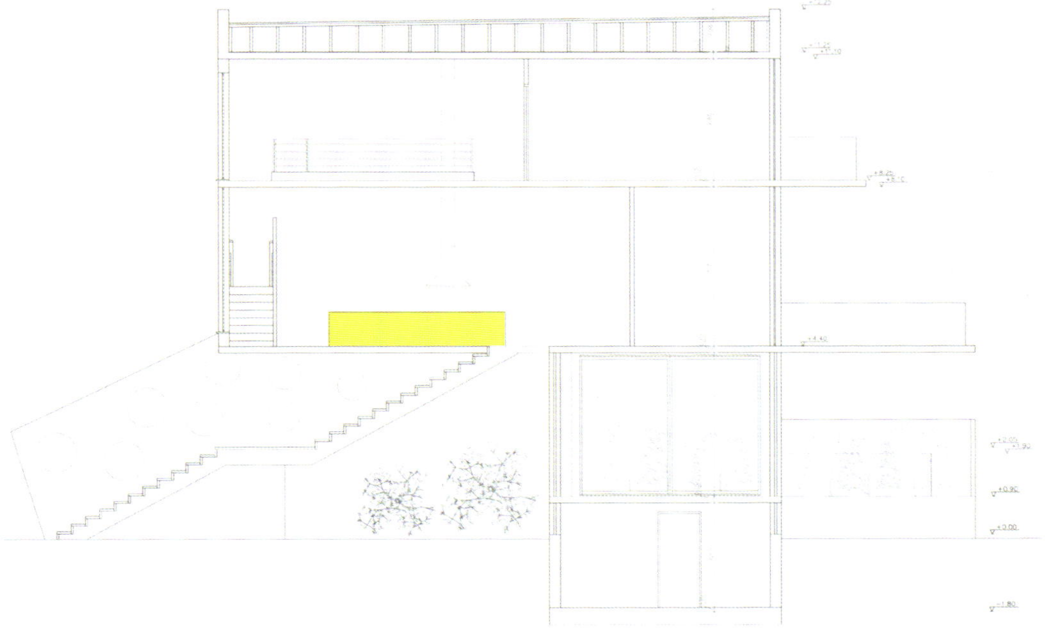

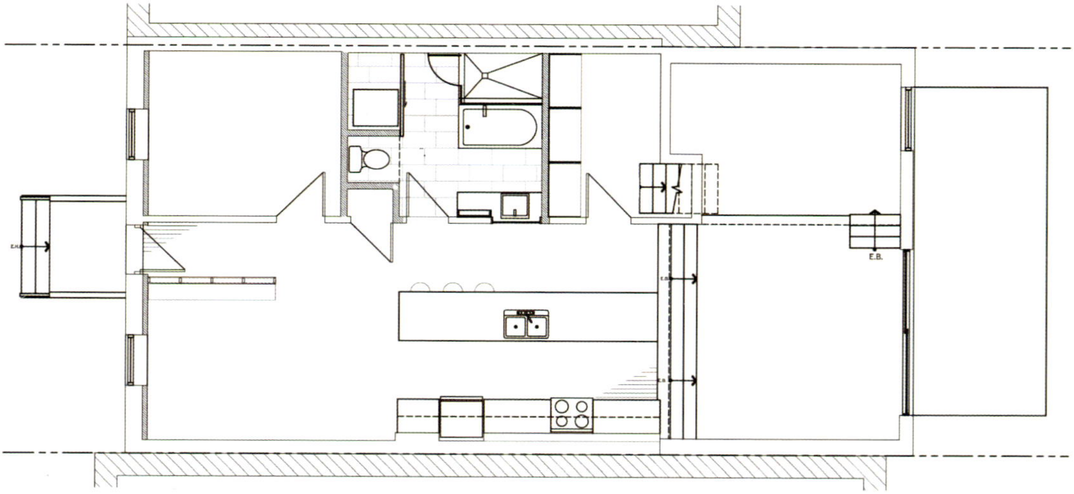
E.B.

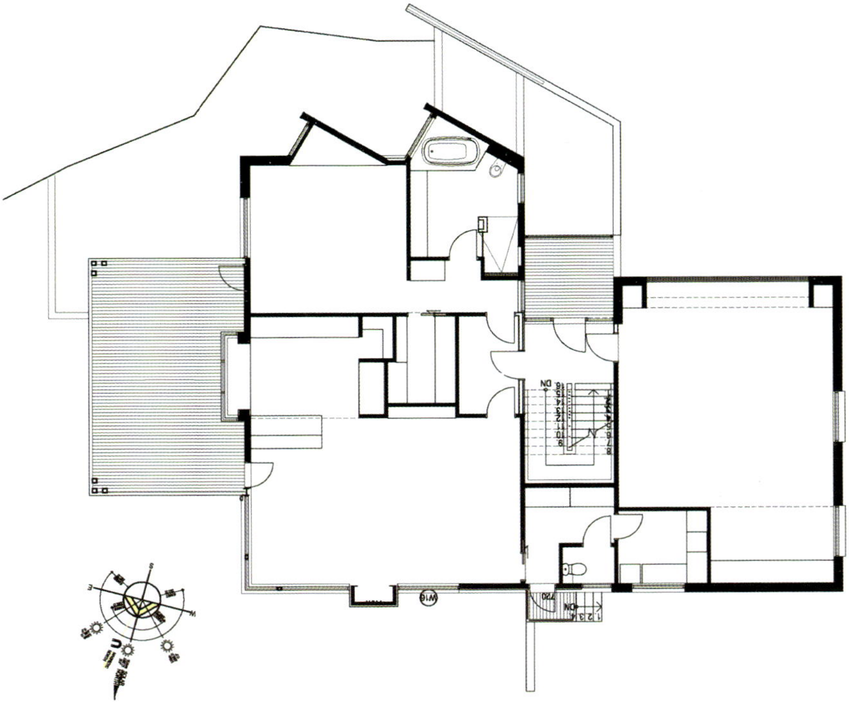

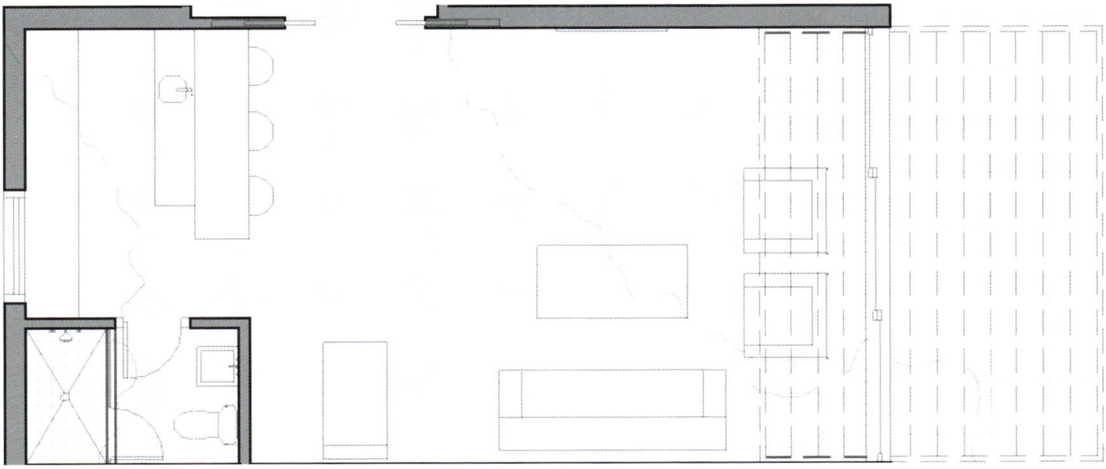

les tomates
rougissent !!!

Les carottes
les tomates rougissent !!!

BOSCH

THE TRAVELER
THE LIMERICK
CLIVE CUSSLER
BLACK HEART
THE TOMMYKNOCKERS
MISCHIEF
People Like Us

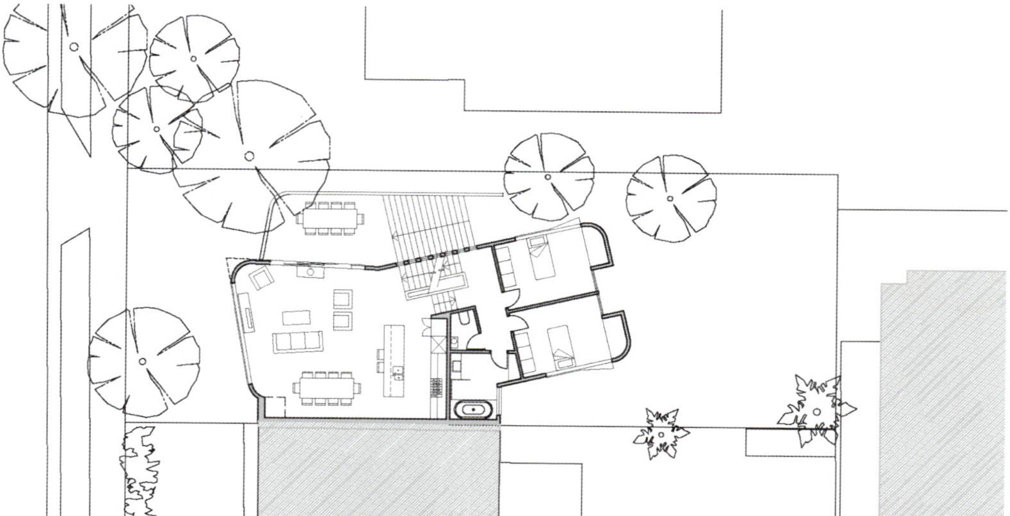

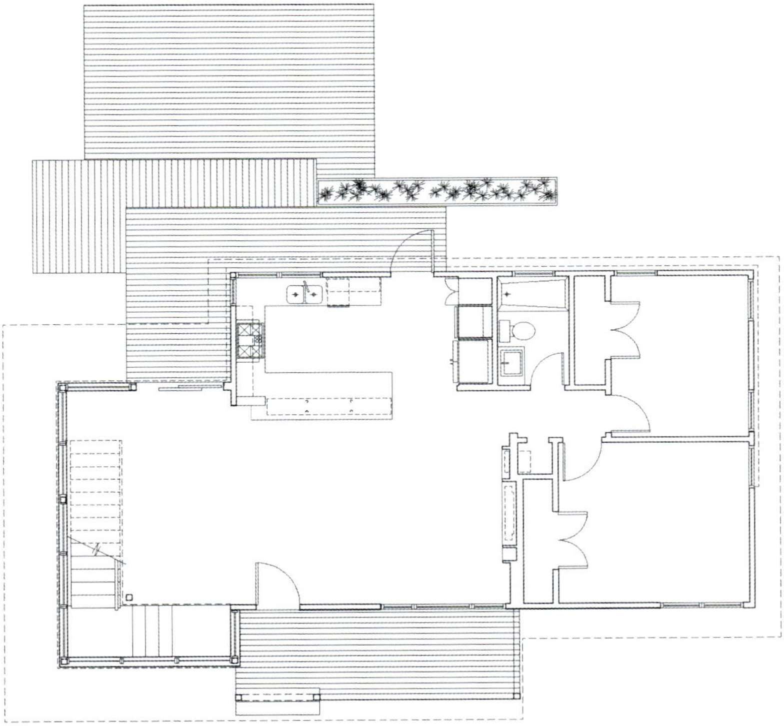

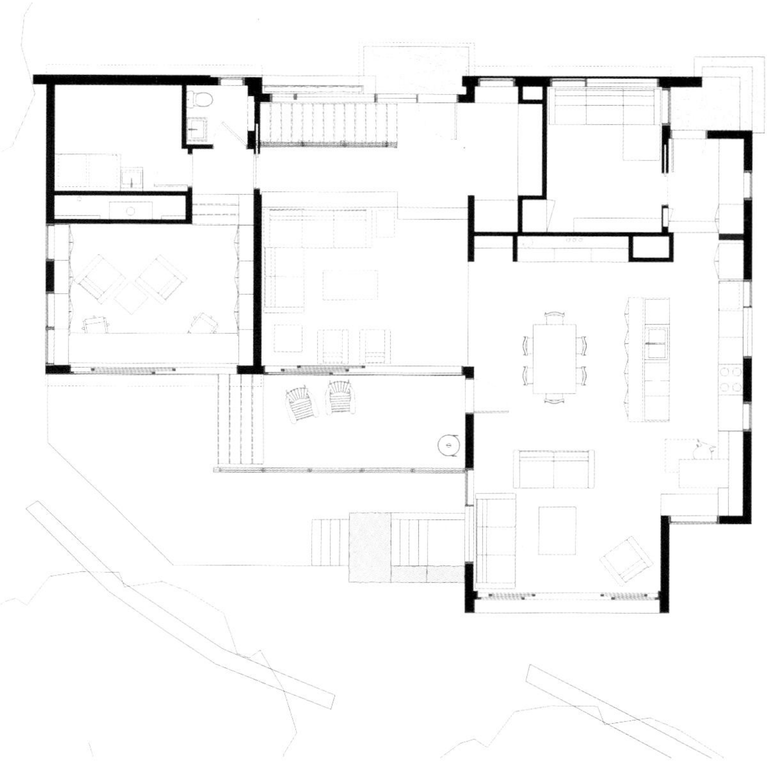

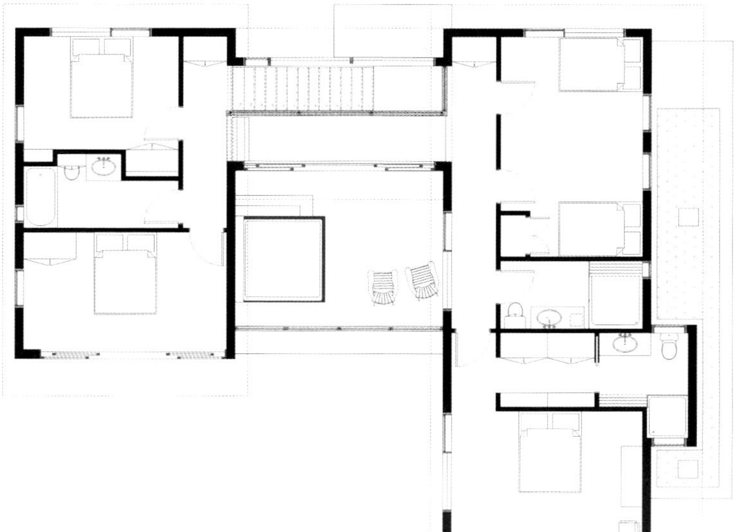

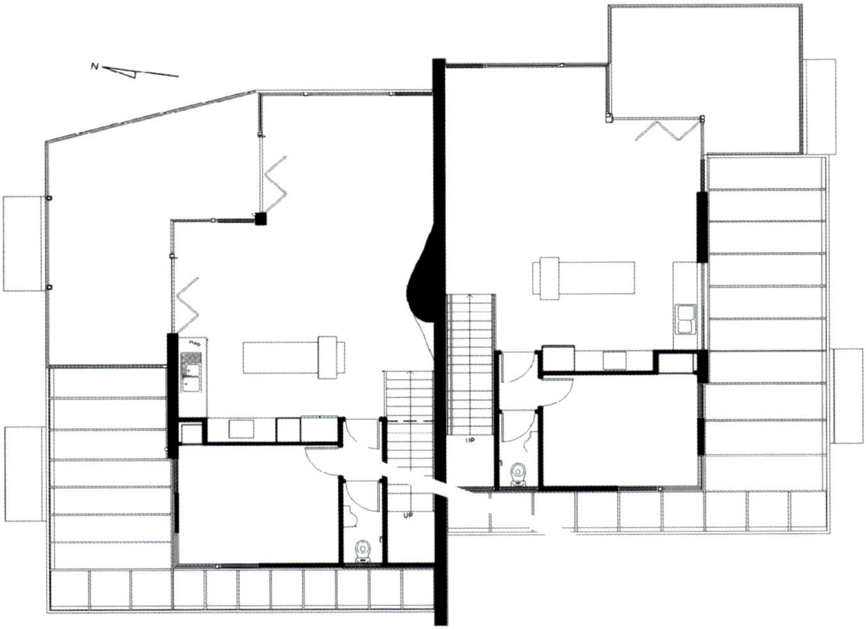
N
UP
UP

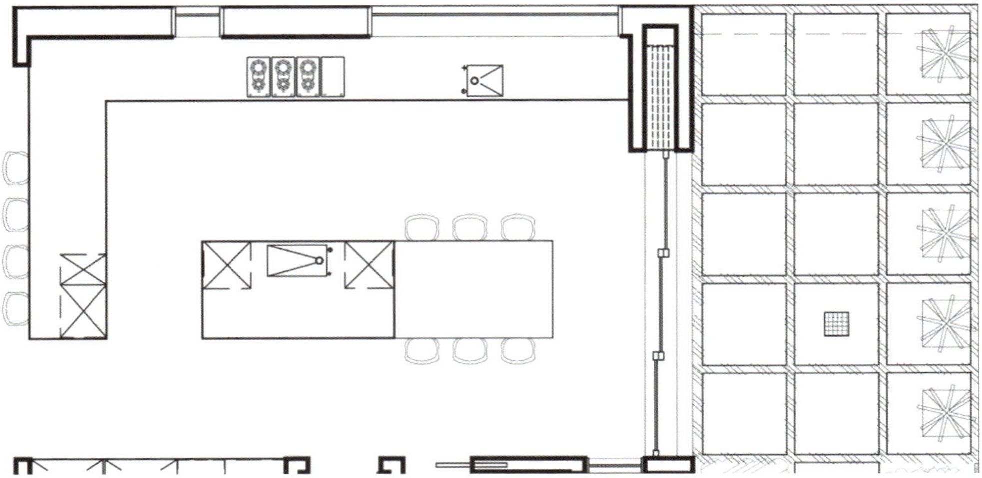

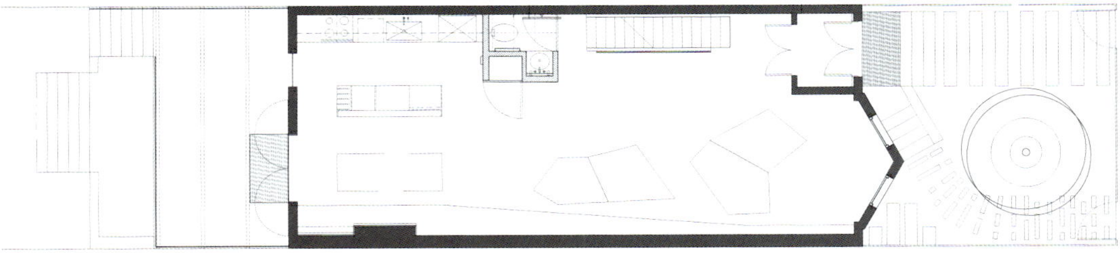

SPOON

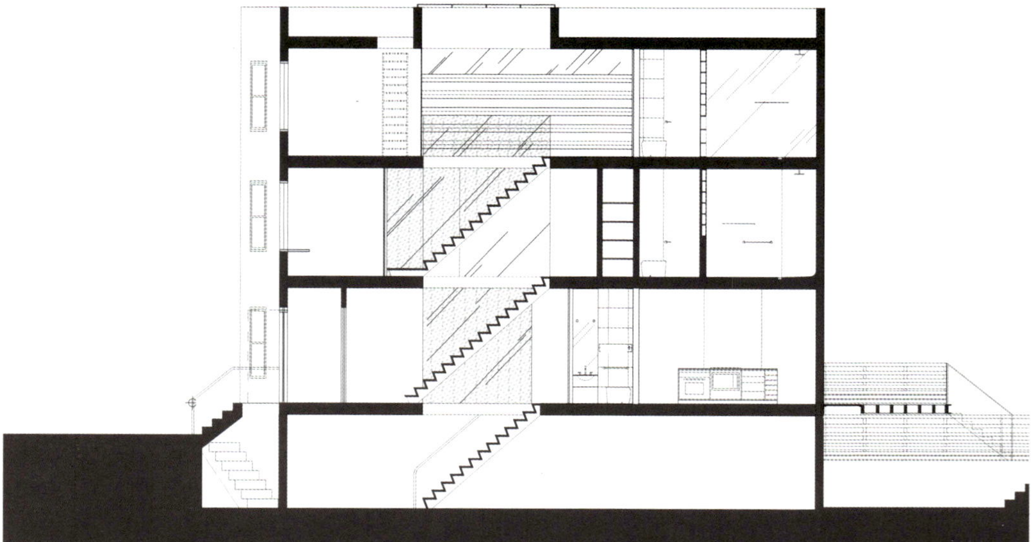

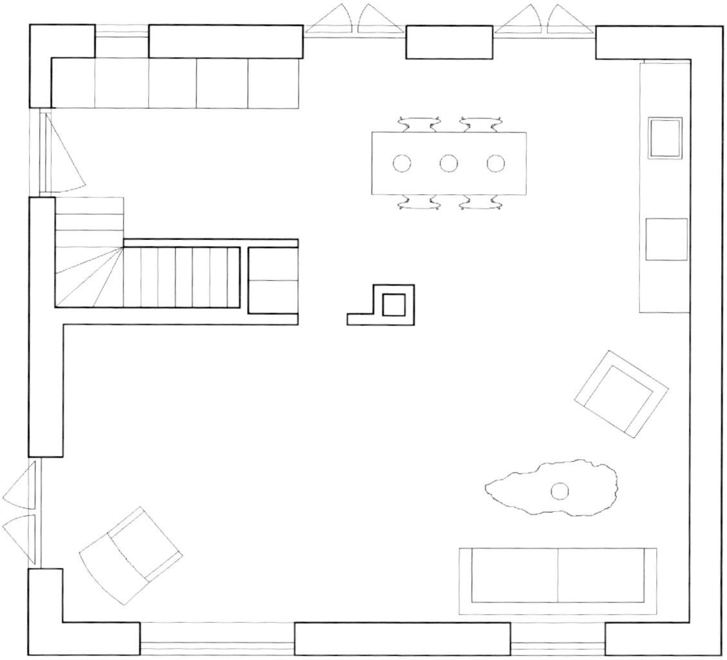

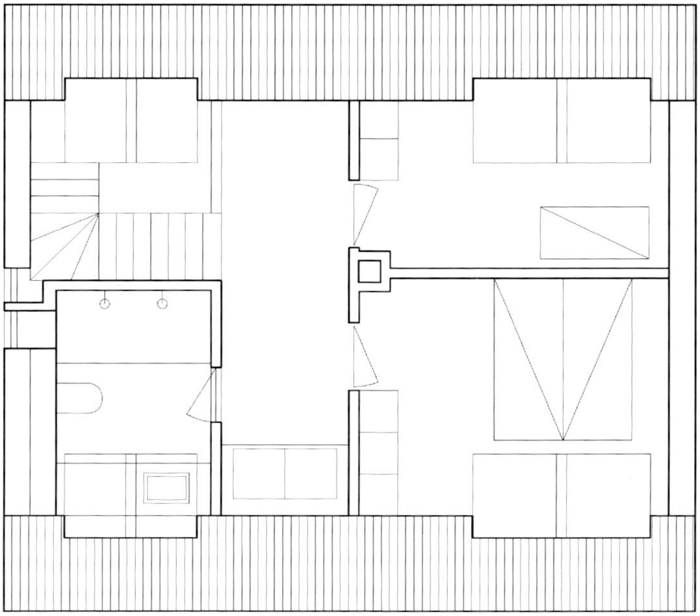

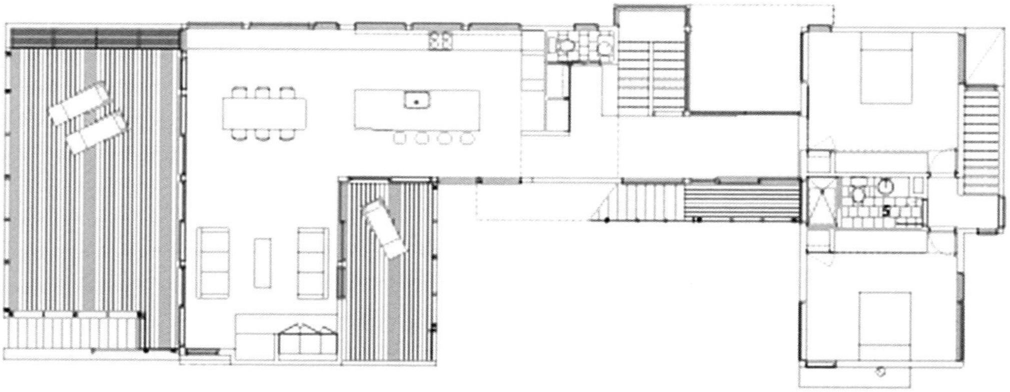

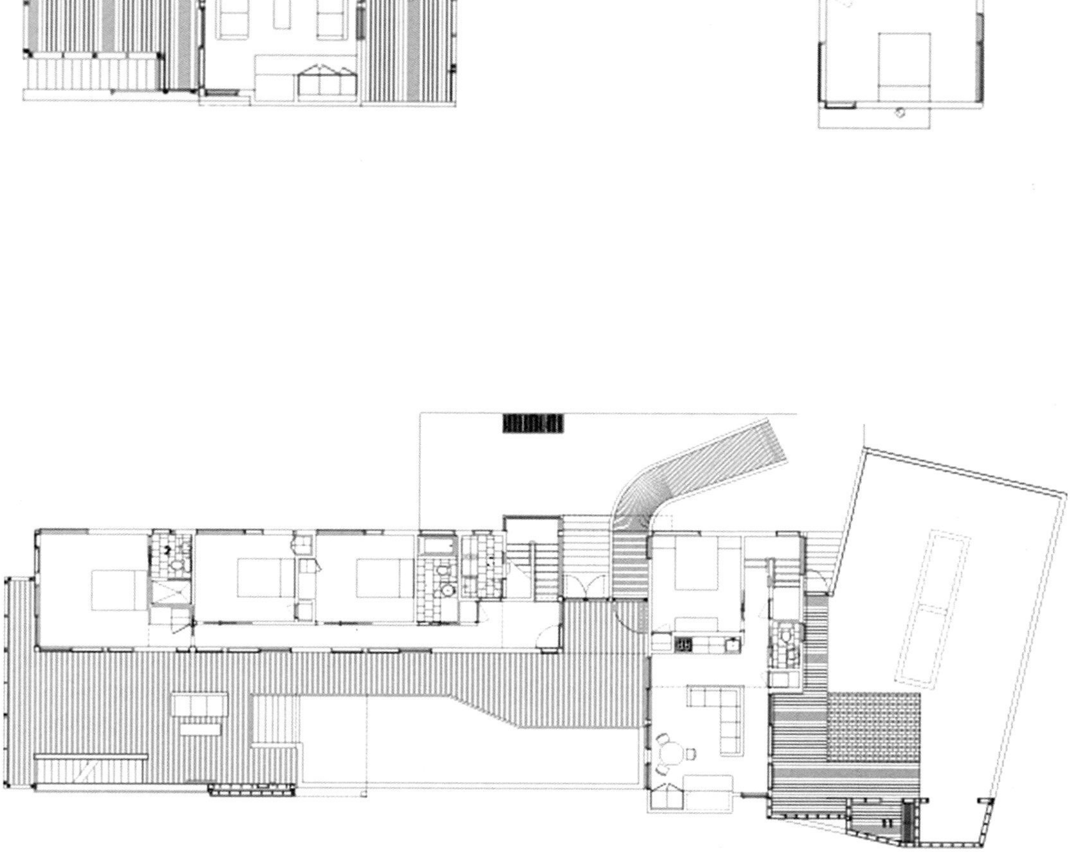

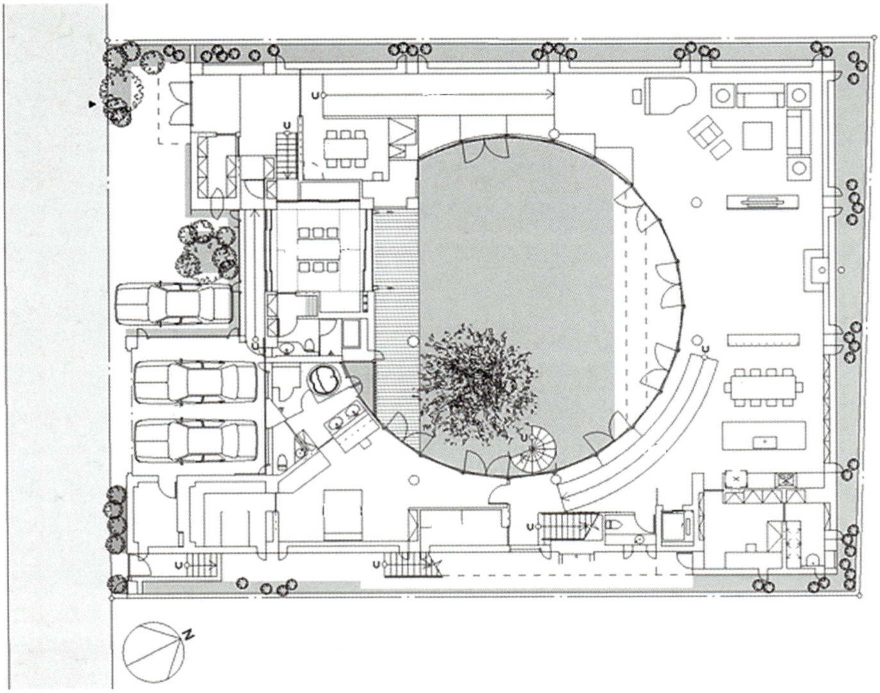
N

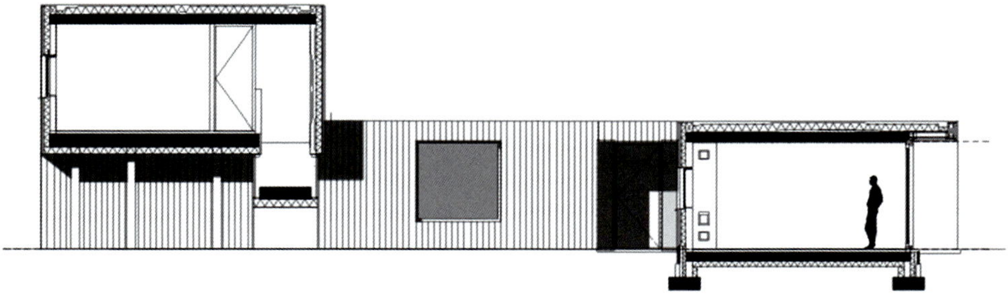

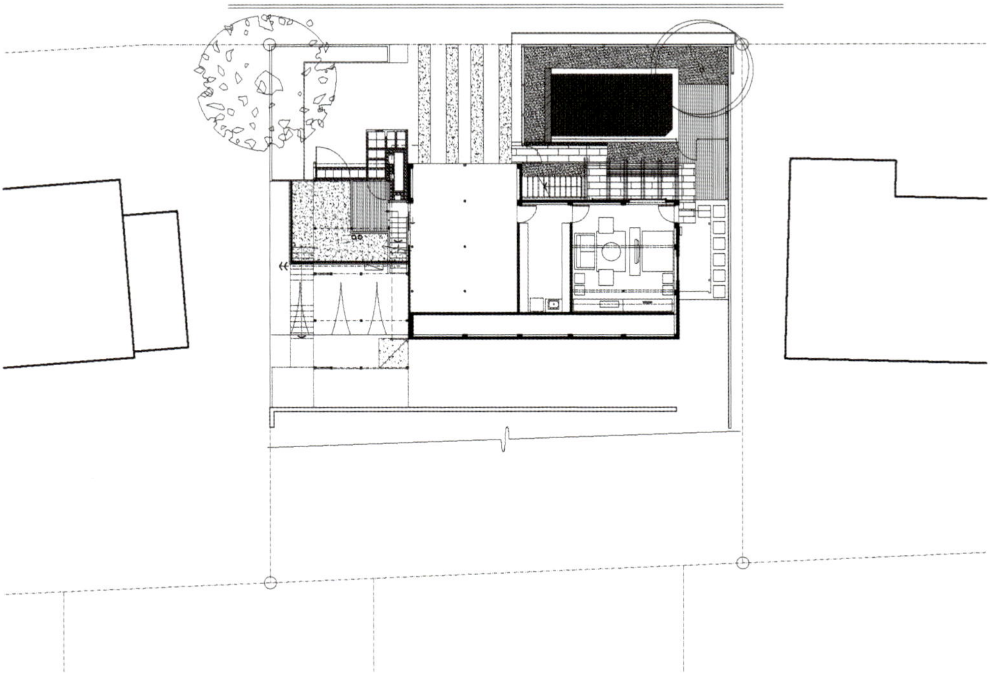

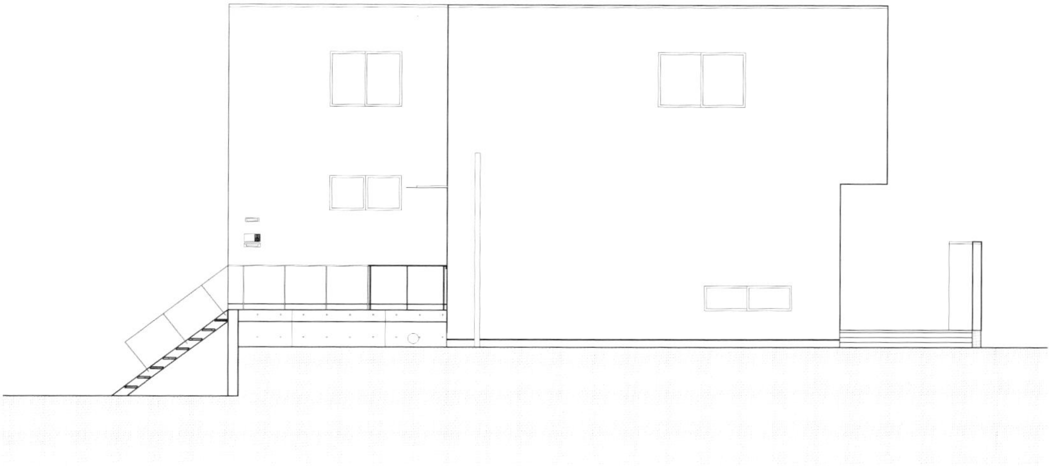

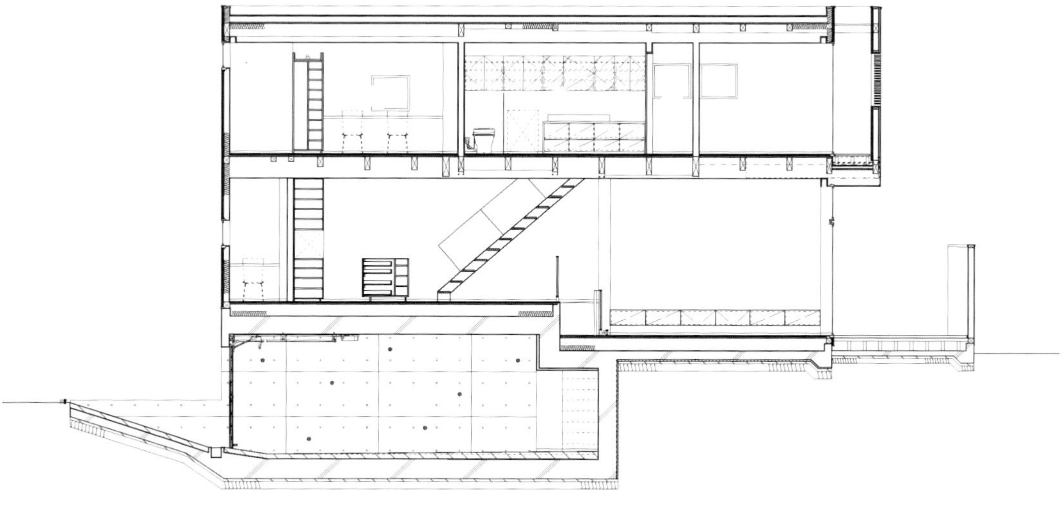

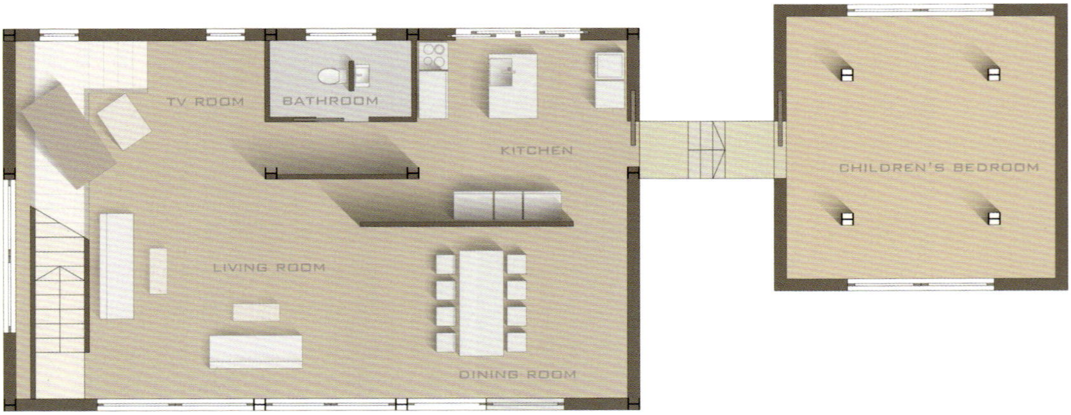
TV ROOM
BATHROOM
KITCHEN
CHILDREN'S BEDROOM
LIVING ROOM
DINING ROOM

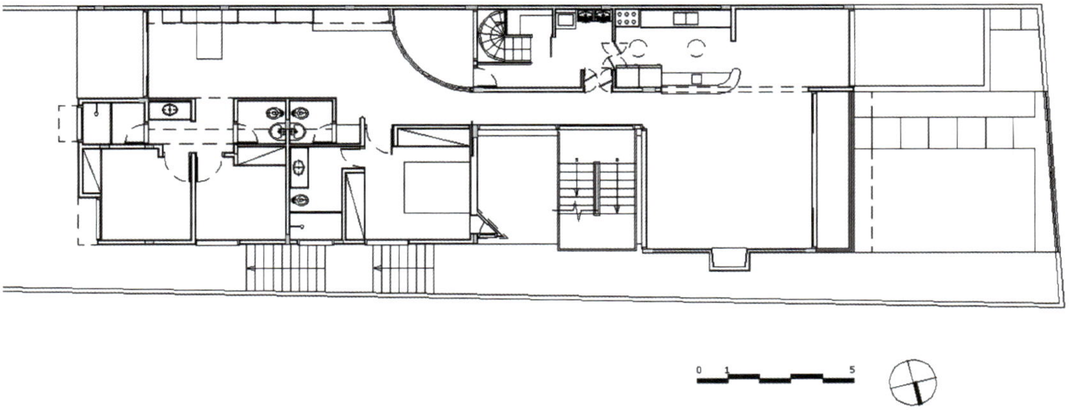
0 1 5

beamer
A
A
beamer screen

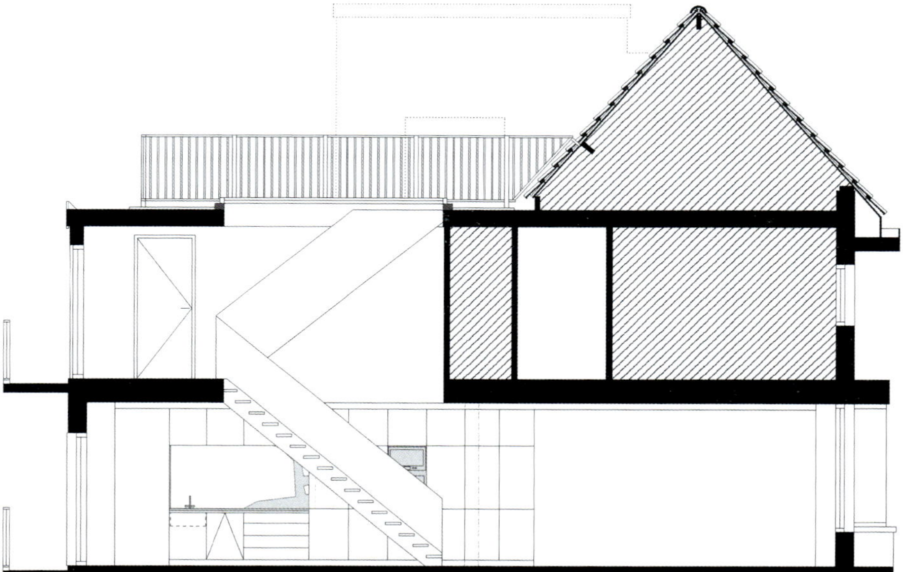

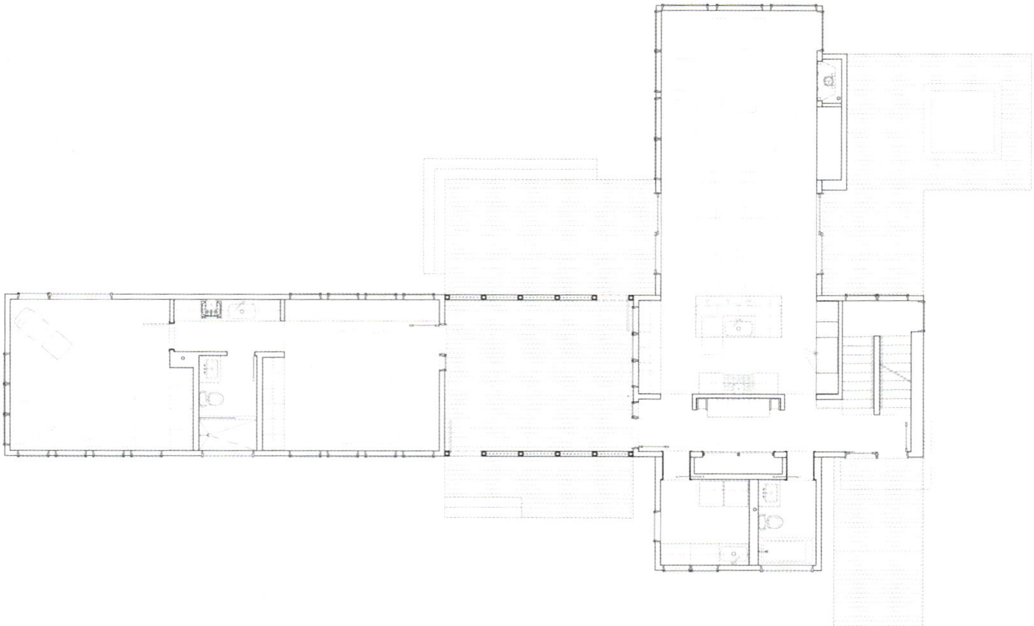

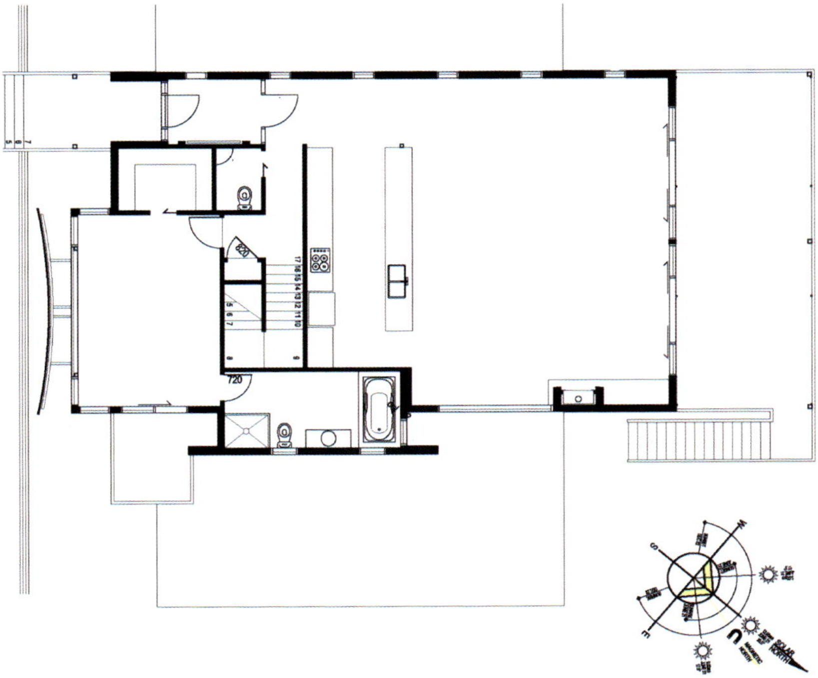

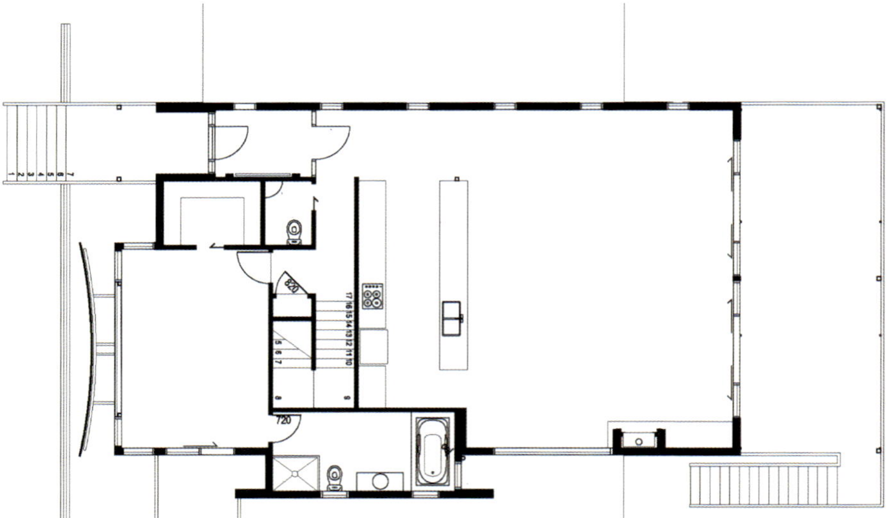
820
720

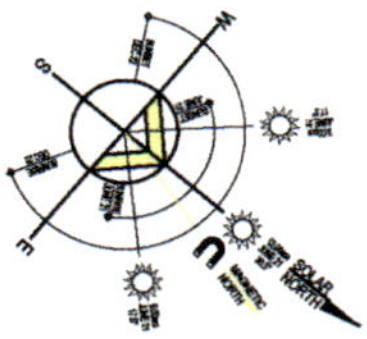
W
S
E
SOLAR NORTH

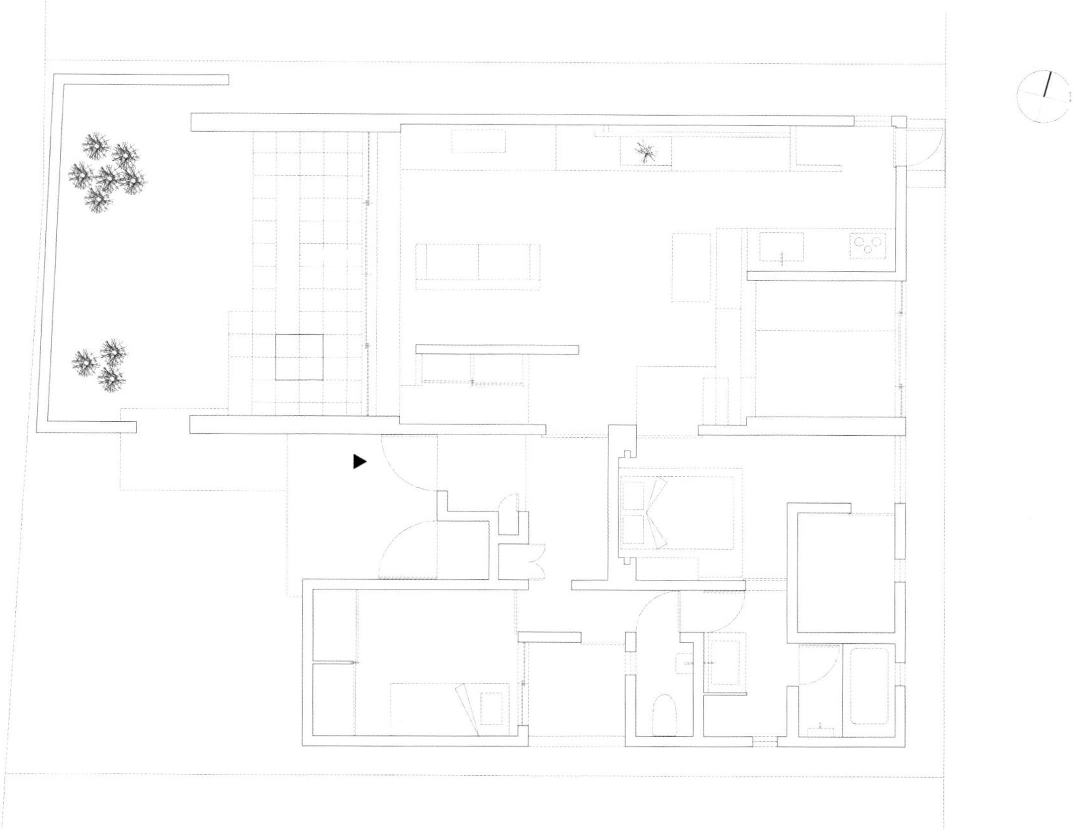

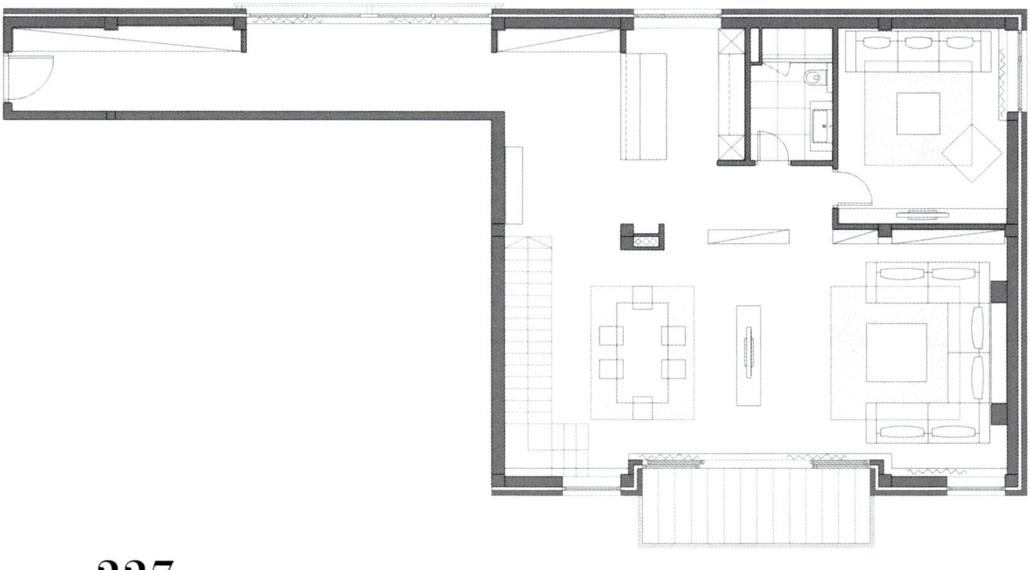

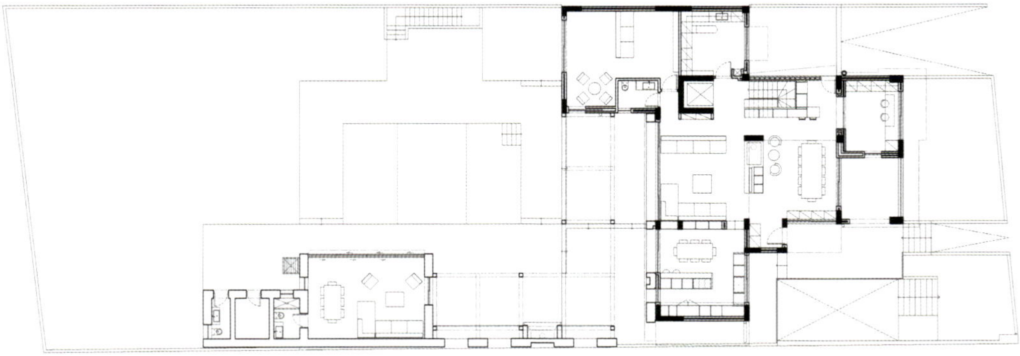

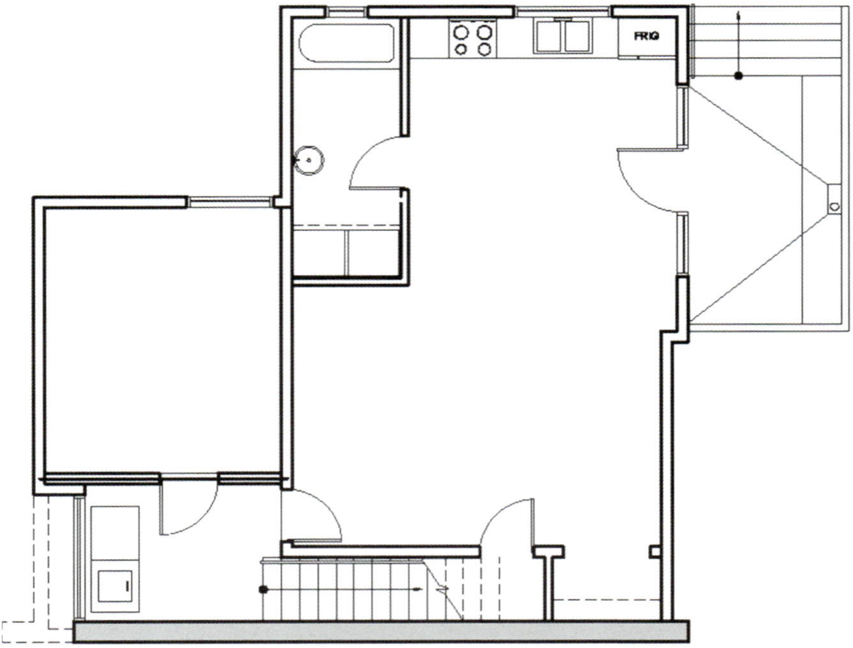
FRIG

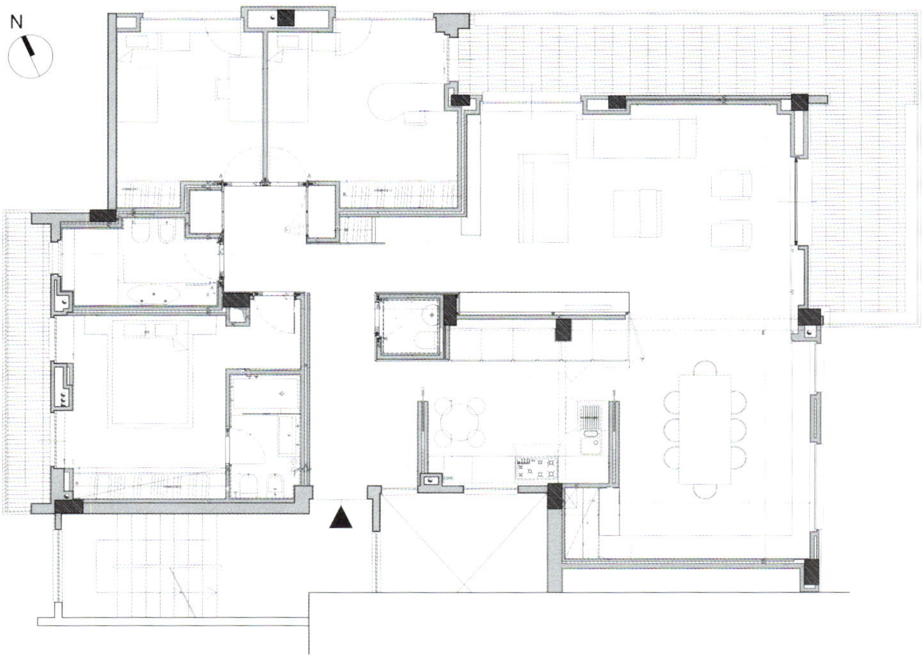
N

Miele

MARTINI

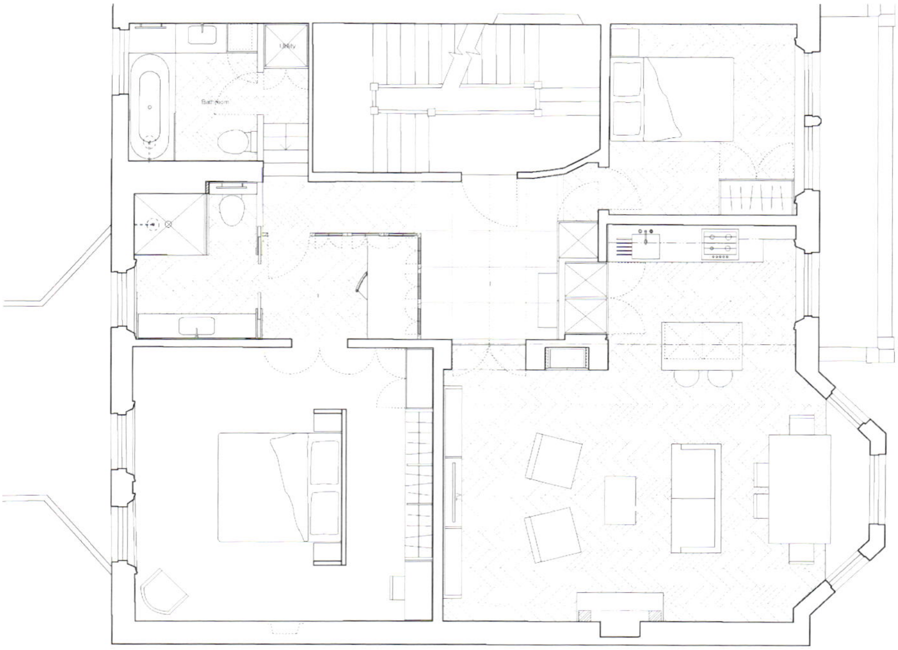
Utility
Bathroom

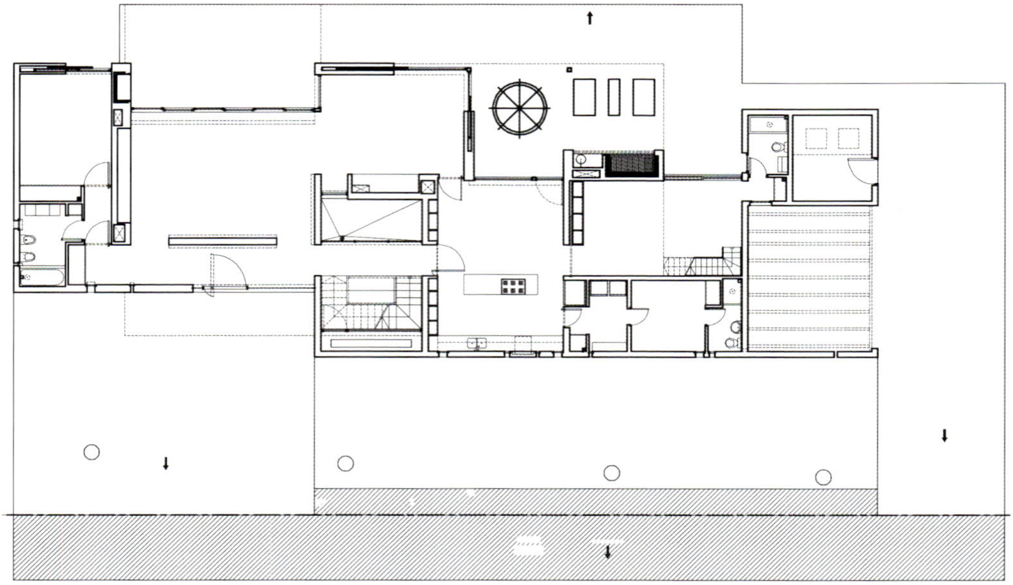

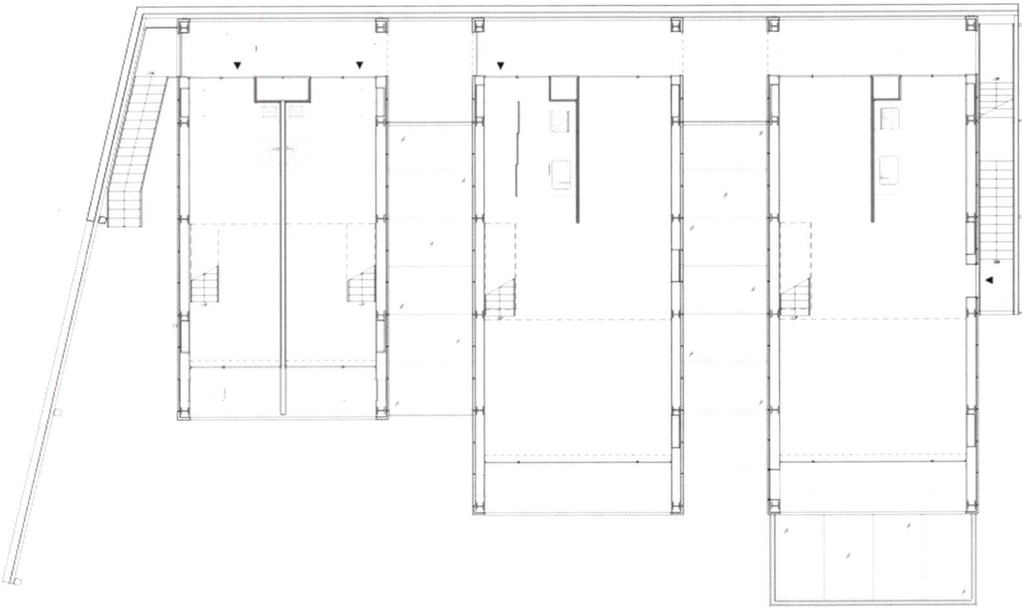

en

ONE-POT
the soup bible

gardening
THE GARDEN DESIGN BOOK
gardening

PASTA

JENN-AIR

VIKING